USING SOLAR FARMS TO FIGHT CLIMATE CHANGE

by Meg Thacher

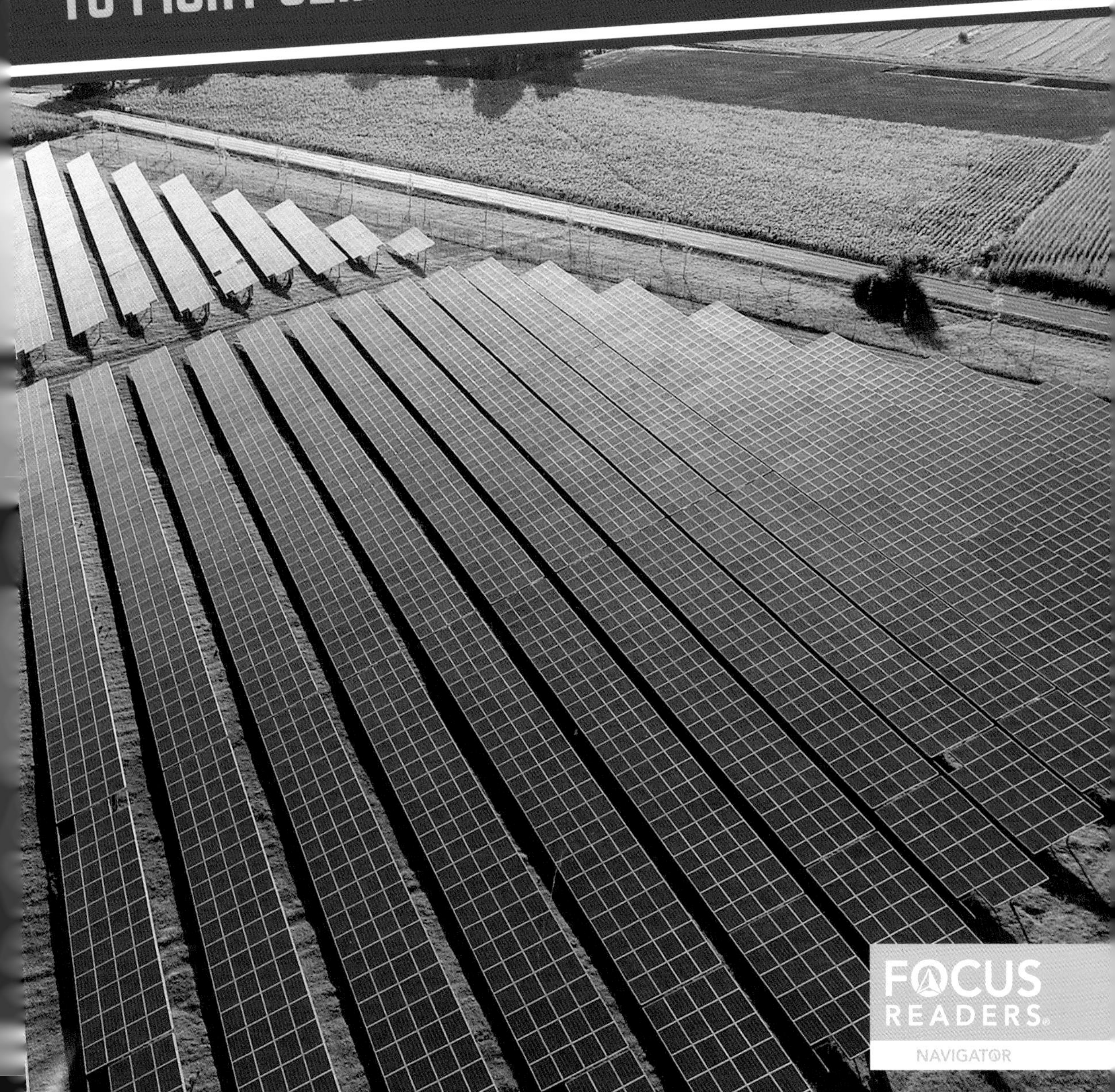

FOCUS READERS
NAVIGATOR

WWW.FOCUSREADERS.COM

Focus Readers is distributed by North Star Editions:
sales@northstareditions.com | 888-417-0195

Produced for Focus Readers by Red Line Editorial.

Content Consultant: Lyra Dumdum Rakusin, Renewable Energy Lecturer at North Carolina State University, College of Natural Resources

Photographs ©: Shutterstock Images, cover, 1, 10, 17 (solar cell), 17 (solar farm), 19, 21, 22–23, 25, 29; Sajjad Hussain/AFP/Getty Images, 4–5; Planet Observer/Newscom, 7; Goddard/GSFC/NASA, 8–9; Red Line Editorial, 13; Goddard/SDO/GSFC/NASA, 14–15; U.S. Department of Energy/Flickr, 27

Library of Congress Cataloging-in-Publication Data
Library of Congress Cataloging-in-Publication Data is available on the Library of Congress website.

ISBN
978-1-63739-277-5 (hardcover)
978-1-63739-329-1 (paperback)
978-1-63739-428-1 (ebook pdf)
978-1-63739-381-9 (hosted ebook)

Printed in the United States of America
Mankato, MN
082022

ABOUT THE AUTHOR

Meg Thacher is the senior laboratory instructor for the astronomy department at Smith College in Massachusetts. She is the author of *Sky Gazing: A Guide to the Moon, Sun, Planets, Stars, Eclipses, Constellations*, as well as 30 children's magazine articles.

TABLE OF CONTENTS

VSPL

CHAPTER 1

THE LARGEST SOLAR FARM

In northwest India, solar panels stand in long, straight rows. By day, the panels soak up sunlight. By night, robots sweep dust off the panels.

These rows of panels are part of Bhadla Solar Park. This solar farm is located in a desert in India. The area is sunny more than 300 days per year. Less than

Workers help at Bhadla Solar Park in the state of Rajasthan, India.

4.5 inches (11 cm) of rain falls each year. As a result, it is the perfect place for a solar farm.

A solar farm is a large group of solar panels. As of 2021, Bhadla Solar Park was the largest solar farm in the world. It covered 62 square miles (160 sq km). It produced enough power for 4.5 million homes.

Solar farms are connected to the electrical grid. The grid is a complex network. It delivers electricity from power sources to users.

Solar farms generate electricity from sunlight. Using them creates no air pollution. For this reason, solar farms

In 2021, Bhadla Solar Park was nearly the size of Washington, DC.

can help fight **climate change**. Countries such as India were helping lead the fight. Bhadla was just one of several large solar farms in India.

CHAPTER 2

FUELING THE CRISIS

Energy is the ability to create heat or move matter. It comes in different forms. For example, moving objects have energy. Light is energy, too. Energy can also change from one form to another.

People in wealthy countries use lots of energy. They use cars and planes to travel. They heat and cool

Electric lighting can be seen from space.

In the early 2020s, approximately 20 percent of all energy that people used went to produce electricity.

their homes with gas and electricity. Electricity powers lights and computers, too. It also powers many factories. These factories make the goods people buy.

Electricity comes from power plants. Most power plants burn fossil fuels. Coal and natural gas are common fossil fuels. Burning these fuels creates heat.

This heat boils water to make steam. The steam spins **turbines**. Turbines turn mechanical energy into electricity.

Burning fossil fuels releases carbon dioxide (CO_2). This **greenhouse gas** enters the atmosphere. It traps heat from

HOW FOSSIL FUELS FORM

Oil and natural gas formed from tiny ocean organisms. These organisms died long ago and sank to the ocean floor. They were buried under layers of sand. The sand became rock. The dead organisms rotted. In contrast, coal formed mainly from plants. These plants died in swampy forests. They sank down through the swamp water. Over millions of years, pressure and heat underground turned all these long-dead organisms into fossil fuels.

the sun. This raises Earth's temperature and causes climate change.

By the 2020s, climate change had become a crisis. Sea levels were rising. Glaciers were melting. Storms were getting stronger. Many areas were becoming hotter and drier. Those places faced more extreme wildfires. Plants and animals were dying out. Many places were becoming unsafe for people to live.

The United States is most responsible for this crisis. Between 1800 and 2020, it produced approximately 25 percent of all CO_2. European countries also played a big role. And China became the largest CO_2 producer in the 2000s. Meanwhile, countries in Africa produced little CO_2. In fact, people in many African countries did not have access to enough energy.

CO_2 IN THE ATMOSPHERE

CO_2 levels are measured in parts per million (ppm).

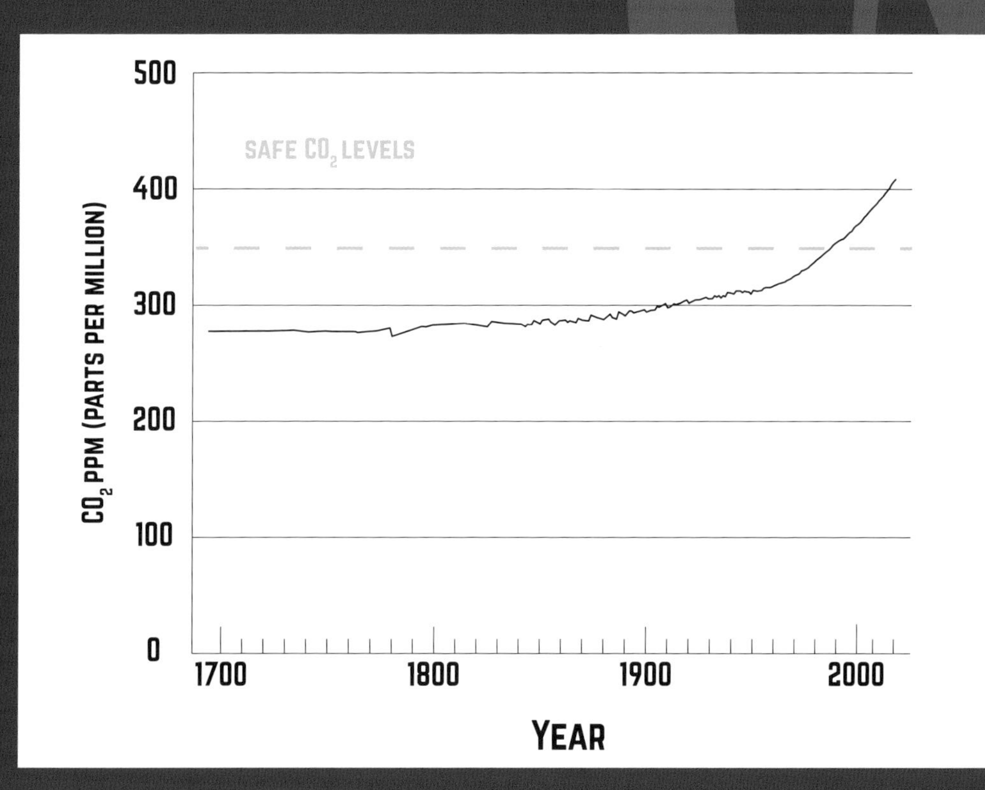

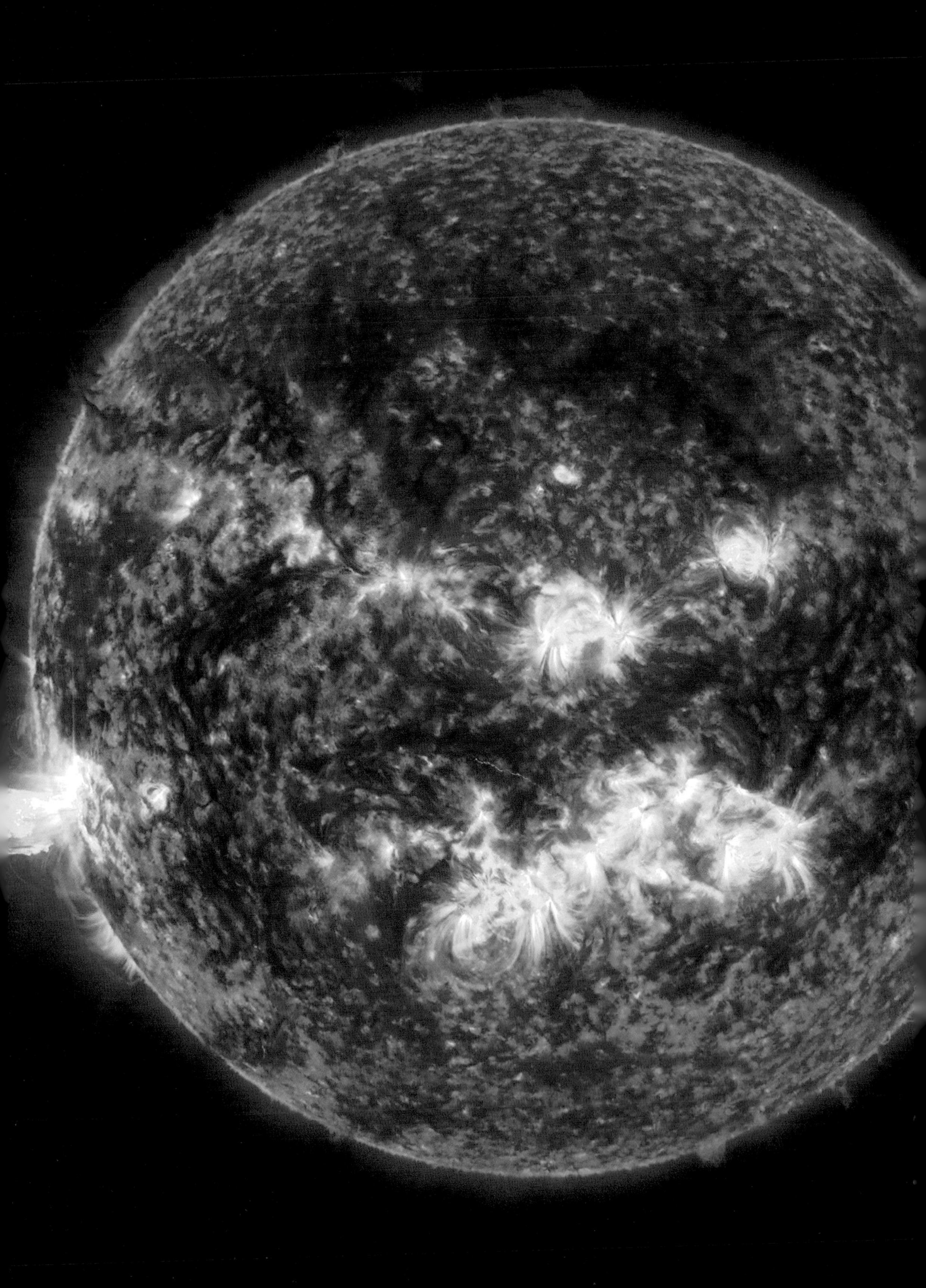

CHAPTER 3

THE SCIENCE OF SOLAR FARMS

The sun creates lots of energy every second. This energy travels to Earth as **photons**. Solar cells capture some of these photons. They turn the photons into electricity. Producing electricity from light is called photovoltaics.

Most solar cells have two layers of **silicon**. One layer has a negative charge.

Unlike fossil fuels, solar energy is renewable. The sun will keep making energy for billions of years.

Its silicon has extra **electrons**. The other layer has a positive charge. It has room for more electrons.

Sunlight hits the silicon. The photons' energy knocks some electrons free. Those particles leave the positive layer. They move to the negative layer. Then metal wires capture the free electrons. These electrons flow through the wires. This flow creates electricity.

One solar cell produces a tiny amount of energy. But many cells can join in a solar panel. Metal connections link the solar cells together. Other materials protect the cells. These include glass and plastic. A solar panel makes useful energy. One solar panel can power several electronic devices.

On a solar farm, thousands of solar panels work together. Cables connect

HOW SOLAR CELLS WORK

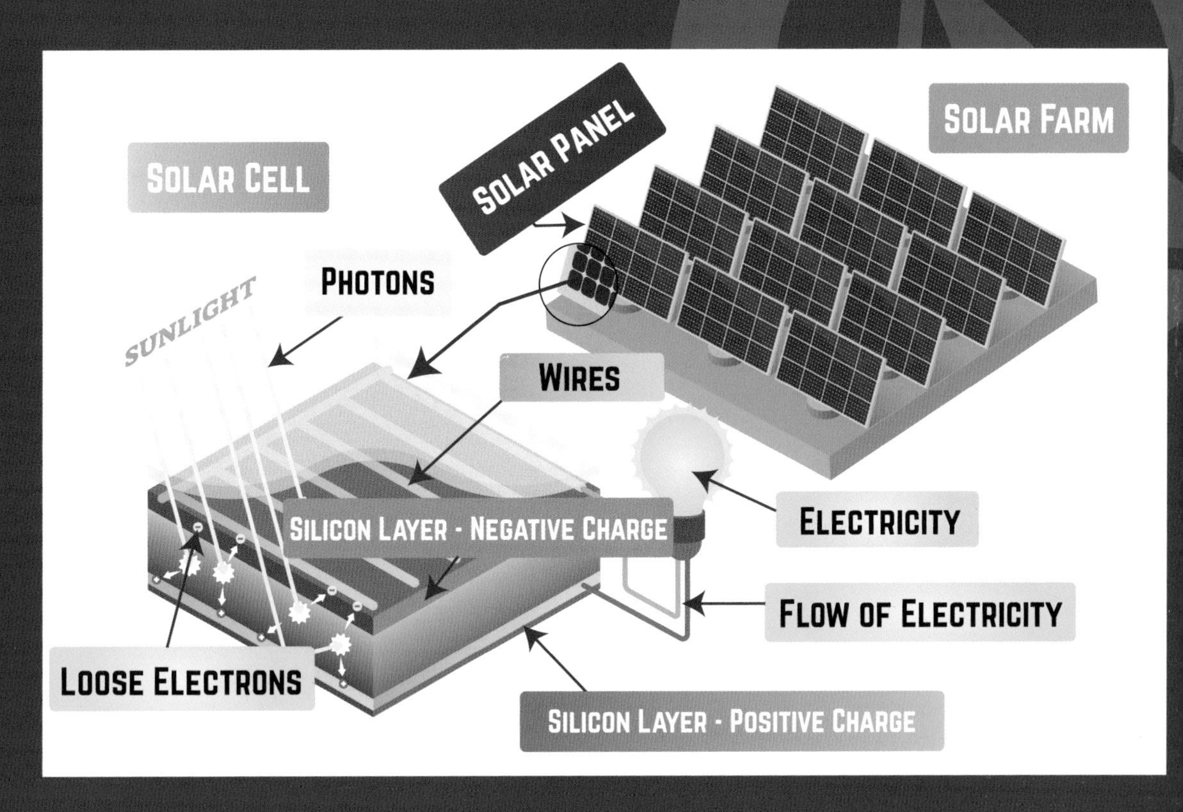

the solar panels. They gather the panels' electricity. That electricity often flows to an electrical grid. Some solar farms can power entire cities.

Solar farms produce some greenhouse gases. For instance, people must get materials to make solar panels. Getting

COLD, SUNNY MORNINGS

People might think solar panels work best during hot, sunny days. They'd be only half right. Sunny conditions are important. But cold temperatures are best. That's because cold electrons have low energy. When light hits them, they gain much more energy than they had before. The difference in energy matters. It helps a solar cell produce more power. Solar panels work best on cold, sunny mornings.

Solar panels have no moving parts. They don't break down as often as turbines. Solar panels last decades.

those materials takes energy. Building a solar farm takes energy, too. However, a solar farm produces far less greenhouse gases than a fossil fuel power plant. That's because using solar farms produces no greenhouse gases. For this reason, solar farms can replace many fossil fuel power plants.

CONCENTRATED SOLAR POWER

A concentrated solar power (CSP) plant does not use solar panels. It has many mirrors surrounding a tower. The mirrors track the sun during the day. They reflect sunlight up to the tower.

The sunlight heats liquid salt in the tower. This liquid runs through pipes to heat a water tank. The water turns to steam. The steam spins a turbine to make electricity.

The first CSP plants were built in the second half of the 1900s. They were not as **efficient** as solar panels. But today's CSP plants are much more efficient.

Solar farms have one very big problem. They can't generate electricity when the sun isn't shining. CSP plants do not have this problem.

In the early 2020s, most concentrated solar power came from Spain and the United States.

The water and salt tanks can store heat. So, they can produce electricity even when the sun is not shining.

CHAPTER 4

THE LIGHT AHEAD

The 2010s saw huge growth in solar power. Even so, solar power still had a long way to go. It provided slightly more than 1 percent of the world's energy. Scientists are working hard to improve the technology.

Reliability was one of solar power's biggest challenges. On cold, sunny

In 2010, the world's solar panels could power nearly 30 million homes. By 2020, solar panels could power 570 million homes.

mornings, solar panels can provide more energy than people need. At night, they don't provide any energy. But people use energy at all times of the day.

For this reason, scientists are working on storing extra solar energy. That way, people can use stored energy when the sun isn't shining.

One way to store energy is **pumped hydropower**. On sunny days, a solar farm delivers needed electricity to users. Then it delivers any leftover energy to a hydropower plant. This energy pumps water into a human-made lake. The lake connects to a river below. At night, the water flows from the lake down into the

In 2022, Europe's largest pumped hydropower plant was in Spain.

river. The water passes through a turbine. The turbine spins, producing electricity.

Other solar farms store their extra energy in batteries. In the early 2020s, batteries for solar power were still costly.

But their costs are going down. More solar farms have batteries for energy storage.

Scientists are improving solar power in other ways, too. Some are developing more efficient solar cells. For example,

SUN TRACKERS

Solar panels work best when sunlight hits them at a certain angle. But the sun moves across the sky during the day. So, solar farms use solar trackers. A solar tracker is a system of computers and motors. It points solar panels at the sun throughout the day. In this way, the panels can receive more of the sun's energy. Solar trackers use energy. But on solar farms, it is worth it. Trackers help solar farms collect up to 20 percent more energy.

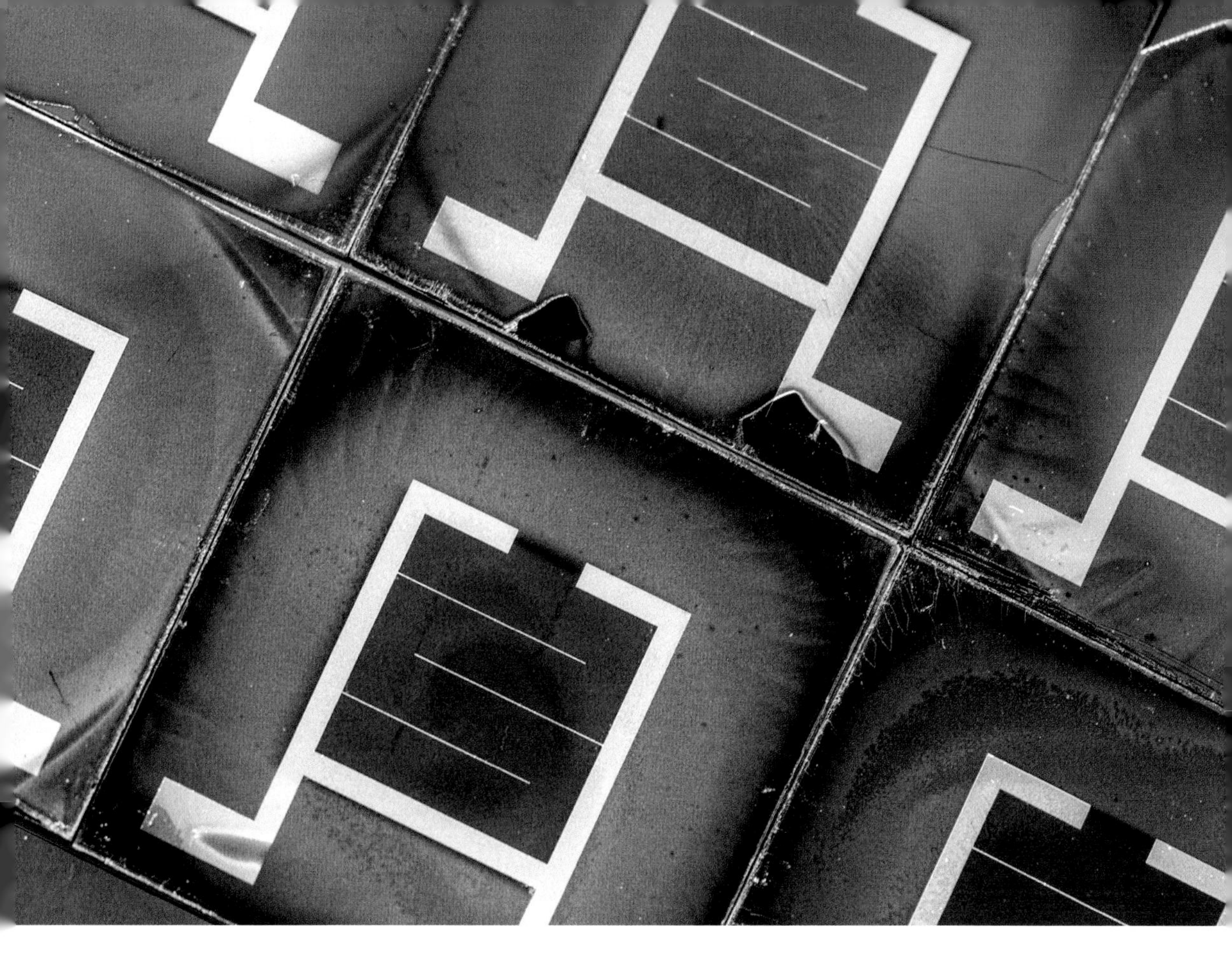

In 2020, solar cells with silicon and perovskite were 6 percent more efficient than cells with only silicon.

silicon cells cannot use most of the sun's energy. Scientists are testing other materials. One is perovskite. It is more efficient, but it is costly. However, adding a thin layer of perovskite could help.

Silicon can have its own environmental challenges. The material comes from rock and sand. It must be separated from the rock. Factories use heat and chemicals to do so. Factory workers must be very careful to get rid of these chemicals safely. Plus, dust from silicon can make people very sick.

In response, scientists found ways to reduce the demand for solar panel materials. One solution was recycling old solar panels. Old panels often still have usable material. Also, broken solar panels can be repaired before they're recycled. These panels can be reused for low-power needs. They can last more than 25 years.

In the early 2020s, Vietnam was a leading country in building solar farms.

By the early 2020s, photovoltaic energy was the cheapest source of electricity. And solar power was continuing to get cheaper and more efficient. The climate crisis was already severe. But solar power was ready to help in an important way.

FOCUS ON

USING SOLAR FARMS

Write your answers on a separate piece of paper.

1. Write a letter describing the main ideas of Chapter 2.
2. Would you want the area where you live to be powered by a solar farm? Why or why not?
3. Which material is part of most solar cells?

   A. silicon
   B. liquid salt
   C. perovskite
4. What can happen on cold, sunny mornings?

   A. Solar farms can stop producing electricity.
   B. Solar farms can produce more electricity than needed.
   C. Solar farms can start turning electricity into photons.

Answer key on page 32.

GLOSSARY

climate change

A human-caused global crisis involving long-term changes in Earth's temperature and weather patterns.

efficient

Accomplishing as much as possible with as little effort or as few resources as possible.

electrons

Charged particles that can be in atoms or on their own.

greenhouse gas

A gas that traps heat in Earth's atmosphere, causing climate change.

photons

Particles of light energy.

pumped hydropower

Relating to the generation and storage of power using water.

silicon

The second most common element in Earth's crust. It can conduct electricity in some situations.

turbines

Rotating machines that are turned by water, steam, or air to produce power.

TO LEARN MORE

BOOKS

Hardyman, Robyn. *Solar Power.* Bridgnorth, England: Cheriton Children's Books, 2022.

London, Martha. *Stopping Climate Change.* Minneapolis: Abdo Publishing, 2021.

MacCarald, Clara. *Energy from the Sun.* Lake Elmo, MN: Focus Readers, 2022.

NOTE TO EDUCATORS

Visit **www.focusreaders.com** to find lesson plans, activities, links, and other resources related to this title.

INDEX

Answer Key: 1. Answers will vary; **2.** Answers will vary; **3.** A; **4.** B